Tesla FSD 13.2.1

A Sneak Peek into the Future of Self-Driving, Featuring 26 New Innovations

Joe E. Grayson

Copyright © 2024 Joe E. Grayson, All rights reserved.

No part of this publication may be reproduced, distributed, or transmitted in any form or by any means, including photocopying, recording, or other electronic or mechanical methods, without the prior written permission of the publisher, except in the case of brief quotations embodied in critical reviews and certain other noncommercial uses permitted by copyright law.

Table of Contents

Table of Contents

Introduction

Chapter 1: Understanding Tesla FSD 13.2.1

Chapter 2: The 26 New Features - An In-Depth Breakdown

Chapter 3: How These Features Impact the Future of Self-Driving

Chapter 4: The Bigger Picture - Tesla's Vision for the Future

Conclusion

Introduction

Tesla has always been more than just a car company. Its mission goes far beyond building electric vehicles; it aims to accelerate the world's transition to sustainable energy. Founded in 2003 by a group of engineers including Elon Musk, Tesla set out to challenge the dominance of traditional internal combustion engine vehicles and redefine the automotive landscape. From the release of its first electric roadster to the mass-market Model 3, Tesla's commitment to innovation has consistently pushed the boundaries of what's possible in electric vehicle (EV) technology.

But Tesla isn't just revolutionizing the car industry — it's also transforming the way we think about energy consumption and production. Through its ventures in solar power and energy storage with products like Solar Roofs and Powerwalls, Tesla aims to create a holistic ecosystem that reduces humanity's reliance on fossil fuels. With the introduction of their electric vehicles, Tesla has not only proven that sustainable transportation can be high-performance, but has also demonstrated that EVs can be luxurious, stylish, and practical.

Tesla has taken a bold and progressive approach with its vehicles, especially when it comes to autonomy. While most car manufacturers are just beginning to dabble in autonomous driving technologies, Tesla has already integrated self-driving capabilities into its cars, setting the stage for what could be a future where driving itself becomes a thing of the past. Their Autopilot system, which has continually evolved, is already capable of handling much of the driving tasks in specific situations, but the real game-changer lies in Tesla's Full Self-Driving (FSD) package.

The importance of FSD cannot be overstated. Full Self-Driving is the cornerstone of Tesla's vision to bring autonomous vehicles to the mainstream. Through advanced software updates, Tesla is not just enhancing its vehicles' capabilities; it's actively working toward making its cars fully autonomous. FSD is designed to handle a broad range of tasks, from automatic lane changes to navigating complex intersections and even driving in city environments. Over time, it's expected to evolve to the point where the car will require little to no human intervention.

One of the most remarkable aspects of Tesla's approach is its emphasis on over-the-air software updates. Tesla owners don't have to wait for a new model release to get the latest features or improvements. Instead, the company can push updates directly to the car, ensuring that its vehicles stay at the cutting edge without requiring a visit to the dealership. This method allows for continuous improvements and refinements, enhancing both performance and functionality with each update.

The 2024 Holiday Update is a prime example of Tesla's commitment to enhancing the user experience. Each year, Tesla releases a major software update during the holiday season, and 2024's update comes with a staggering 26 new features. For Tesla owners, these updates represent not just new functionalities, but a significant leap forward in terms of safety, convenience, and overall driving experience. For those already accustomed to the idea of continuous software updates, it's clear that Tesla isn't just adding new features—it's reshaping what a car can do. These updates aren't just about keeping up with the competition; they are about maintaining Tesla's leadership in the industry by pushing the envelope on what's possible.

Tesla's ability to push out software updates on a regular basis allows the company to constantly improve its vehicles in ways that most car manufacturers can't match. Traditional car makers often have to wait for new model releases to implement changes, but Tesla's approach ensures that every car on the road can benefit from the latest technological advancements. Whether it's a small improvement in battery efficiency or a major upgrade to the FSD system, these updates keep Tesla vehicles at the forefront of the industry.

For owners, this means their Tesla vehicle becomes more than just a car—it becomes a dynamic and ever-evolving product. Every year, Tesla's Holiday Update has become an eagerly anticipated event, offering owners a glimpse into the future of driving. With new features that enhance safety, entertainment, and driving comfort, the update reinforces Tesla's reputation for continually improving the driving experience.

The 2024 Holiday Update, with its wealth of new features, is a perfect example of how Tesla uses regular software updates to reshape the driving experience and cement its role as a leader in the future of automotive technology.

Chapter 1: Understanding Tesla FSD 13.2.1

Tesla's journey toward self-driving technology has been nothing short of revolutionary. It all began with the introduction of **Autopilot**, a semi-autonomous driving system that made its debut in 2015. While it was never intended to be a fully autonomous driving solution, Autopilot was a critical first step in what would become a broader vision of full autonomy. At the core of Autopilot was the idea of assisted driving—allowing the vehicle to perform tasks like steering, braking, and acceleration, but always with the expectation that the driver would remain engaged and ready to take control if necessary.

In 2016, Tesla introduced a more advanced version of Autopilot, capable of navigating on highways and changing lanes with minimal input from the driver. It was a leap forward, but still, the system required the driver's full attention. Tesla continuously improved its software, pushing out updates that added new capabilities, such as adaptive cruise control and automatic lane changes. This iterative process of improving the system through over-the-air updates laid the foundation for what would later be known as **Full Self-Driving (FSD)**.

By 2019, Tesla's self-driving technology reached a new milestone with the introduction of FSD Beta. FSD was designed to take over more complex driving tasks, such as navigating city streets, recognizing stop signs, traffic lights, and pedestrians, and eventually driving itself through intersections and urban environments. This was a game-changer for Tesla owners, as the FSD system moved closer to the long-awaited goal of full autonomy, albeit with a cautious approach—drivers were still expected to supervise the system's operation, as it wasn't fully autonomous at that stage.

The road to true autonomy, defined as **Level 5**, where the car drives itself in any situation with no human intervention, is a complex one. Tesla's path to this goal has not been without its challenges. The biggest hurdle has been creating an AI system capable of interpreting the vast array of real-world driving scenarios that a human driver encounters every day. Unlike traditional vehicles with simple sensors, Tesla relies on a network of cameras and radar to provide a 360-degree view of the environment around the car. The challenge is not just in collecting data but in processing that data in real-time to make decisions that a human driver would normally make instinctively.

To accelerate the development of FSD, Tesla has turned to its massive fleet of vehicles on the road, which collect data and provide feedback that the company uses to train its AI systems. This real-world data has been invaluable in improving Tesla's self-driving capabilities and in refining its neural networks. Every time a Tesla car on the road encounters a new situation—be it a complex intersection, a pedestrian crossing unexpectedly, or a car cutting into its lane—the data is fed back into Tesla's systems, helping the AI to "learn" and improve. This constant feedback loop is a major differentiator for Tesla compared to traditional automotive manufacturers, many of which rely on simulations and less real-world data.

The development of Tesla's Full Self-Driving system hasn't been a straight path. While Tesla has made significant progress, it has also encountered its share of setbacks, including regulatory challenges, safety concerns, and technical limitations. Despite these obstacles, the company has continued to refine its software and make improvements, all while maintaining its vision of creating an autonomous vehicle that could safely navigate the world without human intervention.

Tesla FSD 13.2.1 represents a key point in this ongoing evolution. As Tesla's latest release in the Full Self-Driving suite, version 13.2.1 includes a range of improvements designed to enhance the performance and capabilities of the system in real-world driving conditions. Unlike earlier versions of FSD, which were often limited to highway driving or required significant human supervision, FSD 13.2.1 represents a step closer to achieving full autonomy.

One of the key features of **FSD 13.2.1** is its **improved handling of complex urban environments**. Tesla vehicles with this update are now better equipped to navigate through city streets, dealing with a wider variety of traffic situations. Whether it's interpreting ambiguous road signs, managing unpredictable pedestrian movements, or responding to sudden changes in traffic flow, FSD 13.2.1 is designed to make these tasks easier and more reliable. This is a significant improvement, as navigating urban areas is one of the most challenging aspects of self-driving technology.

Another important upgrade in version 13.2.1 is the **enhanced response to stop signs and traffic signals**. Tesla cars with FSD now recognize these signals more accurately, and they've become better at deciding when to yield, stop, or proceed. The system also does a better job of interpreting complex intersections, which is crucial for making autonomous vehicles a reality in everyday driving situations. Prior versions of FSD struggled in situations with heavy cross-traffic or in non-standard intersections, but 13.2.1 has made strides in improving the system's decision-making process in these contexts.

Perhaps most importantly, **FSD 13.2.1** continues to refine the system's overall **safety and reliability**. While Tesla vehicles are already equipped with advanced safety features, this update makes the system even more robust, helping to prevent potential accidents and ensuring that the car behaves in a predictable and controlled manner in a wider variety of driving conditions. These incremental updates are crucial, as safety is the cornerstone of autonomous driving technology. For Tesla, each update is not just about adding new features, but about refining the system to handle the full complexity of driving, ensuring that when the time comes for full autonomy, it will be as safe as human driving, if not safer.

FSD 13.2.1 doesn't mark the end of Tesla's journey toward full autonomy, but it does represent a significant leap forward. By improving the car's ability to handle complex environments, recognize traffic signals more accurately, and make smarter decisions on the road, Tesla is closing the gap between its vehicles and true self-driving cars. The road to Level 5 autonomy is still long, but each software update, including FSD 13.2.1, brings Tesla closer to realizing its vision of a future where cars drive themselves, safely and efficiently, without human intervention.

Chapter 2: The 26 New Features - An In-Depth Breakdown

The 2024 Tesla Holiday Update introduced an impressive collection of 26 new features that enhance the overall driving experience and continue to push the boundaries of what a car can do. Tesla's commitment to innovation is evident in this update, which brings not only functional improvements but also touches of fun and personalization. These updates are a clear indication that Tesla sees its vehicles as more than just a mode of transportation—they are dynamic, ever-evolving products that grow with their owners' needs.

Among the standout features of the update is the integration of the **Tesla app with the Apple Watch**, making it easier than ever for owners to stay connected to their vehicle. This new capability allows Tesla owners to control key vehicle functions directly from their wrist, providing an added layer of convenience. Whether it's unlocking the car, checking the battery status, or even controlling the climate, all of this can now be done seamlessly from the Apple Watch. For busy Tesla owners who are always on the go, this feature offers a level of accessibility that simply wasn't available before, reducing the need to pull out a phone or reach for the car key.

Another notable enhancement is the ability to **save Dash Cam and Sentry Mode clips directly to your phone**. This feature provides a quick and easy way for Tesla owners to store footage captured by the car's cameras, which can be invaluable in case of an incident or for reviewing unusual events. Whether it's footage from an accident, a close call, or a curious encounter, the ability to save clips directly to the phone enhances security and peace of mind. For those who've used Tesla's security features in the past, this update streamlines the process, ensuring that important moments are saved instantly without any extra steps. This simple but effective improvement makes it easier to access and share footage when needed.

The **auto-shift feature on stockless Model 3s** is another impressive addition. This update allows the car to automatically shift between drive and reverse without the need for the driver to manually engage the gearshift. This feature is particularly useful for those navigating tight spaces or parking situations. It reduces the mental load on the driver, making everyday driving tasks smoother and more intuitive. Whether you're pulling into a parking spot or reversing out of one, the car's system automatically senses the need to shift, simplifying the process and reducing the chance of user error.

Maintenance summaries and **route search** are two new features that have a profound impact on the convenience of ownership. Maintenance summaries give drivers an easy-to-read snapshot of their vehicle's health, making it simpler to stay on top of necessary repairs and scheduled services. By receiving regular updates about the car's condition, owners can avoid surprises and keep their Tesla running at optimal performance. Similarly, the ability to **search along a route for maintenance services** makes long trips easier to plan, ensuring that drivers can find charging stations or service centers when needed.

The introduction of the **Red Cross Traffic Alert** feature is another vital safety enhancement. This feature alerts Tesla drivers to the presence of emergency vehicles with flashing lights on the road ahead. By notifying drivers of potential hazards before they encounter them, the Red Cross Traffic Alert improves safety by giving drivers more time to react and adjust their driving. It also enhances overall situational awareness, reducing the likelihood of accidents, especially in emergency situations when fast decision-making is critical.

For those who love personalization, the update also includes features designed for the **upcoming Cybertruck**, including **wraps and license plate customization**. The ability to personalize the vehicle's exterior adds a new level of individuality to the already bold Cybertruck. Whether you're looking to make a statement with a custom wrap or simply want to add a unique license plate, Tesla is embracing personalization like never before. Adding a bit of fun to the mix, the **Santa Mode** feature brings seasonal cheer to Tesla owners by transforming the car into a festive mode with playful animations and sounds, making the holidays even more enjoyable.

Tesla continues to redefine in-car entertainment with features like the ability to **schedule light shows** through the Tesla app. This fun, yet practical feature allows owners to synchronize their car's lighting to create a spectacular visual display. Perfect for the holiday season or any special occasion, this update turns your Tesla into a personalized light show, adding an extra layer of customization and entertainment for both owners and onlookers. This feature enhances the idea that Tesla vehicles are not just cars—they are expressions of personality and creativity.

Speaking of entertainment, Tesla has also integrated the **Boomerang game** into the vehicle's infotainment system. This game, along with other entertainment options, adds a playful dimension to the Tesla driving experience. Whether you're waiting for the car to charge at a Supercharger or simply want to entertain your passengers during a road trip, the Boomerang game offers a fun distraction. It's part of Tesla's broader strategy to redefine the in-car experience, turning what might otherwise be idle time into moments of enjoyment.

The **rear-seat control of passenger seats** is a feature designed with convenience and comfort in mind. Now, passengers in the back of the car can adjust the front seats using the rear touch screen. This feature is particularly useful for families or individuals who frequently ride in the back seat, making it easier to adjust the seating position without needing the driver's assistance. It's a small but significant addition that enhances the overall comfort of the ride for everyone in the vehicle.

Another significant addition is the ability to **play video content while driving**. While this feature is primarily designed for passenger entertainment, it opens up new possibilities for long trips or daily commutes. However, this feature is subject to ethical considerations and safety concerns, as it could potentially distract drivers if misused. Tesla has put safeguards in place to ensure that this feature is used appropriately, but it highlights the evolving nature of in-car entertainment.

Finally, **Supercharger arrival notifications** are a simple but effective enhancement. With this feature, Tesla drivers will be notified when they arrive at a Supercharger station if any charging stalls are out of service. This saves time and frustration, as it provides real-time information about charging availability, allowing drivers to adjust their plans and avoid wasting time at a station that may not have open stalls. This update is a perfect example of how small features can have a big impact on the overall user experience, making the process of owning and driving a Tesla even more seamless.

Each of these 26 features introduced in the 2024 Holiday Update contributes to a richer, safer, and more enjoyable driving experience. Whether through improving convenience, enhancing safety, or adding a touch of fun and personalization, Tesla continues to push the envelope in redefining what a vehicle can be. These updates not only keep Tesla at the forefront of the electric vehicle industry, but they also demonstrate how Tesla is fundamentally changing the relationship between drivers and their cars.

Chapter 3: How These Features Impact the Future of Self-Driving

The evolution of self-driving technology has always been about more than just the hardware in the car. It's the software—the algorithms, the artificial intelligence, the constant learning and improving—that truly shapes the future of autonomous driving. Tesla's approach to self-driving has set it apart from traditional automakers, and one of the key factors that has allowed Tesla to push the boundaries of what's possible is its reliance on **continuous software updates**. These over-the-air updates enable Tesla to improve the performance of its vehicles in real-time, making significant strides in autonomous capabilities without requiring new physical hardware or even a trip to the dealership.

For Tesla, the idea of over-the-air updates is foundational to its self-driving mission. Unlike other manufacturers who may take years to release new models with improved features, Tesla is able to rapidly deploy new software to its existing fleet of vehicles. This means that Tesla owners are continuously receiving the latest advancements in autonomous driving technology. Every update is a chance to refine the car's ability to make intelligent decisions on the road, learn from real-world driving data, and improve the car's understanding of its environment.

Each software update builds on the previous one, making incremental but important progress toward the ultimate goal of **full**

autonomy—where the car can drive itself safely in any situation without human intervention. Tesla's ability to learn from its massive fleet of vehicles is one of the factors that accelerates its path toward autonomy. Every mile driven by a Tesla car feeds into a vast network of data, which is then analyzed and used to improve the vehicle's ability to navigate complex roadways, respond to unexpected events, and make decisions on the fly. This data-driven approach helps Tesla to address challenges that other automakers, who rely more on simulation and less on real-world data, are still grappling with.

The significance of these updates cannot be overstated. With each release—such as the **FSD 13.2.1** update, which further refines Tesla's ability to navigate urban environments and interpret traffic signals—the gap between human driving and autonomous driving continues to close. The car is getting better at handling the nuances of real-world driving, and Tesla is steadily chipping away at the barriers that stand between today's semi-autonomous vehicles and the fully autonomous future that many envision. With **features like improved stop sign recognition, better traffic light handling, and the ability to make safer decisions at complex intersections**, these updates make it increasingly feasible for the car to operate without constant human oversight.

However, this journey is not without its hurdles. Tesla continues to face challenges, not just in terms of technology but also in how society perceives and embraces autonomous driving. Full autonomy, or **Level 5 autonomy**, where a car can handle all driving tasks in any condition, remains an elusive goal. But as Tesla continues to update its vehicles and push forward with its advancements in AI and machine learning, each new feature brings the company closer to that goal. In fact, features like **automatic lane changes, enhanced parking assist, and real-time traffic awareness** are all stepping stones that, in combination, form the backbone of a future where human drivers may not be necessary at all.

Safety and Convenience are central to Tesla's strategy in rolling out these updates. Each new feature is designed not only to improve the driving experience but also to enhance safety—one of the most critical aspects of autonomous driving. For instance, the **Red Cross Traffic Alert** feature, which notifies drivers of emergency vehicles, allows them to make better decisions on the road. This type of real-time awareness is key to preventing accidents and improving overall road safety. Similarly, the ability to **save Dash Cam and Sentry Mode footage directly to a phone** adds another layer of protection for Tesla owners, ensuring that critical moments are easily captured and stored.

Maintenance Summaries are another example of how Tesla is focusing on safety and convenience. The ability to track the health of your vehicle in real time ensures that Tesla owners can stay ahead of any potential issues, getting repairs done before small problems become big ones. This proactive approach to car maintenance not only improves safety but also increases the convenience of owning a Tesla. With a car that is constantly updating and improving itself, owners are less likely to be caught off guard by unexpected breakdowns or performance issues.

But while these safety features are critical, they also represent the balance Tesla has to strike between **innovation** and **user convenience**. As Tesla pushes the envelope with self-driving technology, it must also ensure that each update remains user-friendly and does not overwhelm owners with too many options or too much complexity. For example, while the new **auto-shifting feature** in the stockless Model 3 makes driving easier, it doesn't require the user to learn an entirely new way of interacting with the car. Tesla's goal is to integrate these advancements smoothly into the driving experience without sacrificing the simplicity that has made Tesla vehicles so popular in the first place.

Tesla's focus on **driving experience evolution** is clear, and it goes far beyond just adding new features. The continuous improvement of Tesla vehicles via software updates is changing how owners interact with their cars. What was once seen as a car's static features—fixed at the time of purchase—has evolved into a dynamic, constantly improving product. Features that weren't possible when the car was first purchased can be added months or even years down the line, ensuring that Tesla owners have the latest and best technology at their fingertips. Whether it's the ability to **schedule light shows** or the integration with the **Apple Watch**, these updates reshape the driving experience and ensure that it's tailored to the needs of the owner.

The implications of these features go far beyond just Tesla owners. The industry at large is being reshaped by the constant progress of Tesla's software-driven approach. Other automakers are looking at Tesla's success and considering how they can integrate similar features into their own vehicles. The more that Tesla pushes the boundaries of what's possible with software, the more it forces traditional car manufacturers to reconsider their approach to vehicle design and technology. The long-term impact of Tesla's updates could be transformative not just for driving but for the entire automotive industry. As self-driving technology evolves and software-driven vehicles become the norm, the way we think about cars—and the way we use them—will continue to shift.

Tesla's **long-term vision** is to create a future where driving becomes something that happens without the need for human intervention. But for now, the updates Tesla is rolling out make it clear that we're not far off from a time when driving routines will be dramatically different. With each new update, Tesla is nudging us closer to a future where driving no longer requires constant attention and where vehicles can navigate themselves safely through even the most complex environments. It's an exciting glimpse into what the future of transportation could look like—one where the car becomes not just a tool for mobility but a fully autonomous, intelligent partner in getting from one place to another.

Tesla is changing the world, one update at a time, and with each step, it's not just transforming how we drive, but how we live.

Chapter 4: The Bigger Picture - Tesla's Vision for the Future

Tesla's commitment to **continuous improvement** is more than just a business strategy—it's a fundamental aspect of its ethos. From the very beginning, Tesla has positioned itself as a company that constantly evolves, not just through new car models but through software updates that enhance and refine the performance of existing vehicles. This relentless pursuit of improvement is what sets Tesla apart from traditional automakers, who often release new models with incremental changes every year. For Tesla, the focus is on keeping its vehicles at the cutting edge of technology, using **over-the-air software updates** to improve everything from performance to safety features, to user experience.

This culture of continuous improvement has fostered a unique relationship between Tesla and its owners. Unlike traditional cars, which can feel static as time passes, Tesla vehicles are always evolving. The car you buy today isn't just the car you'll own tomorrow—it's the car that will continue to improve over time, often in ways that were unimaginable at the time of purchase. Every update brings new capabilities, fixes, and optimizations, ensuring that Tesla owners are always receiving the best possible driving experience, even as they continue to use a vehicle that's months or years old.

This strategy has given Tesla a significant edge in the competitive **electric vehicle (EV) market**. In an industry where technology is advancing rapidly and customer expectations are growing, staying ahead of the curve is crucial. With traditional automakers racing to catch up in the EV space, Tesla is constantly innovating to stay at the forefront. By offering regular software updates and integrating new features, Tesla ensures that its vehicles maintain their relevance, even as competitors release new EV models. This gives Tesla a considerable advantage in retaining existing customers and attracting new ones, as they know they'll receive not just a car, but a car that improves over time.

As for **full autonomy**, the journey toward **Level 5 autonomy**—where a car can operate without any human intervention—remains an ongoing challenge. While Tesla's Full Self-Driving (FSD) system has made significant strides in recent years, the company is still working toward a fully autonomous driving experience. The release of updates like **FSD 13.2.1** shows just how far Tesla has come, but the road ahead is still long. At present, Tesla's cars can handle a wide range of driving tasks, including **lane changes, stop sign recognition, and traffic light handling**. However, **Level 5 autonomy** would require the vehicle to handle every driving scenario—urban, rural, in any weather condition, and in unpredictable situations—without any input from the driver.

Each new update is a step closer to that goal. Features like **auto-shifting between drive and reverse**, improved traffic navigation, and real-time hazard detection are key components of the broader push toward full autonomy. Tesla is constantly refining its systems through updates, using the data from its fleet of vehicles to improve the software. As more cars on the road provide feedback, Tesla's cars will get smarter, more capable, and more adept at handling complex driving situations. However, there are still many hurdles to overcome, both in terms of technology and regulation. While Tesla is ahead of the pack in terms of technology, the timeline for achieving **Level 5 autonomy** is still uncertain. It could be a few years—or it could take longer—before Tesla and the industry as a whole are able to fully realize this ambitious goal.

Despite the challenges, the introduction of **FSD 13.2.1** and similar updates provide a clearer picture of Tesla's progress. These features indicate that Tesla is well on its way to achieving more complex levels of autonomy. The improvements in **real-time traffic alerts**, **predictive navigation**, and **adaptive decision-making** are all signs that Tesla is continually refining its systems, pushing closer to the day when cars will drive themselves.

Looking to the **future of Tesla vehicles and software**, it's clear that Tesla's mission is to keep improving, innovating, and redefining the driving experience. The company has already changed the automotive landscape by pushing the boundaries of electric vehicle technology and self-driving capabilities. As Tesla continues to refine its FSD system and integrate more advanced features, we can expect even more exciting developments on the horizon.

In the near future, **Tesla vehicles** could feature even more advanced AI, capable of interpreting complex driving environments with greater accuracy and making real-time decisions in ways that today's vehicles simply can't. **Battery technology** could see huge leaps, allowing for longer ranges, faster charging times, and reduced costs. Tesla is also likely to continue expanding its **network of charging stations**, making long-distance electric driving more practical than ever. These advancements will not only benefit Tesla owners but could potentially reshape the entire automotive industry.

The **impact on the automotive industry** as a whole will be profound. Tesla's approach to software updates, data-driven development, and continuous improvement has already changed how we think about cars. Other automakers will likely follow suit, implementing their own versions of over-the-air updates, AI integration, and autonomous driving features. As Tesla leads the charge, the competition in the EV space will intensify, pushing all companies to innovate faster and deliver more advanced vehicles. The emphasis on software and connectivity will likely lead to a shift in how vehicles are built and sold, with much greater focus on the **user experience** and the **digital aspects** of car ownership.

Furthermore, Tesla's commitment to improving its vehicles through software also speaks to a broader trend of **software-first** thinking in the automotive world. As cars become more connected, more autonomous, and more reliant on software to function, the focus will shift from just building hardware to creating systems that can continuously evolve and adapt. In this future, a car is no longer a static machine, but an evolving platform that gets smarter over time. This will likely have far-reaching consequences not just for the automotive industry but for the way we think about mobility, transportation, and technology in general.

Tesla's journey toward full autonomy and its commitment to continuous software updates are not just changing the way we drive—they're shaping the future of the entire industry. As the company moves closer to achieving **Level 5**

autonomy, its updates and innovations will continue to set the standard for what's possible in the world of electric vehicles and autonomous driving. What started as a bold vision for a sustainable future has grown into a transformative force in the automotive world, and the best is yet to come. Tesla is not just driving toward the future; it's actively creating it, one software update at a time.

Conclusion

As we look back on Tesla's **2024 Holiday Update**, it's clear that this software upgrade is more than just an incremental change—it represents a significant leap forward in the world of **self-driving** and **electric vehicles**. The **26 new features** included in this update are not just additions to a car's capabilities but signals of the profound changes that are coming to the way we think about driving and car ownership. From **Apple Watch integration** to **real-time traffic alerts** and even fun additions like the **Santa Mode**, these features demonstrate Tesla's commitment to continually enhancing the driving experience, making it safer, more convenient, and more personalized.

For **Tesla owners**, this update offers a glimpse into a future where driving is not just about getting from point A to point B, but about creating a more connected, enjoyable, and streamlined experience. Features like the ability to save **Dash Cam** footage, **auto-shifting** between drive and reverse on stockless Model 3s, and the **Supercharger arrival notifications** reflect Tesla's understanding that modern drivers demand more than just a car—they want a platform that evolves with their needs. The importance of these updates cannot be overstated, as they not only improve safety and convenience but also push the boundaries of what's possible with software-driven vehicles.

In the context of the broader **self-driving community**, these updates offer a clear indication that we are moving closer to **full autonomy**. While we are not there yet, Tesla's continuous progress in refining its **Full Self-Driving (FSD)** system shows just how far we've come. The improvements in **traffic awareness**, **predictive navigation**, and decision-making capabilities are crucial steps in Tesla's journey toward a future where cars can drive themselves safely in any environment, without human intervention. The updates in **FSD 13.2.1** are just the latest in a long line of improvements, each one bringing us closer to a fully autonomous future.

Tesla's path forward is one of constant innovation. The company has set its sights on more than just selling cars—it is committed to transforming the entire **transportation industry**. With its approach to over-the-air updates, **AI integration**, and **continuous improvements**, Tesla is setting a new standard for what modern vehicles can and should be. The integration of

autonomous features, vehicle customization options, and an improved **in-car entertainment system** is just the beginning. As Tesla refines its software and expands its fleet of vehicles, we can expect even more groundbreaking changes in the way we think about and interact with our cars.

The **road to full autonomy** is still long, but Tesla's progress has been nothing short of remarkable. With each update, the cars are becoming smarter, more capable, and closer to being able to handle all driving tasks without human intervention. While Level 5 autonomy may still be some years away, the strides Tesla has made in this direction have been game-changing. It's not just about the technology; it's about creating a **new era** in which cars are no longer just machines—they are intelligent, autonomous systems that can navigate the world with precision and ease.

Looking ahead, Tesla is poised to continue its role as a **pioneer in the future of transportation**. As it continues to roll out new features, improve existing ones, and push the envelope on what's possible with autonomous driving, Tesla is not just shaping the future of its vehicles but the future of how we move, how we travel, and how we interact with the world around us. Tesla's commitment to **continuous improvement** will ensure that its vehicles remain at the forefront of innovation for years to come. The journey toward a fully autonomous future may take time, but Tesla is on the right path, and its role in the evolution of transportation is one that will be remembered for generations to come.

In conclusion, Tesla's 2024 **Holiday Update** isn't just about 26 new features—it's about a bold vision for the future of driving. Each update brings us one step closer to the world that Elon

Musk and Tesla envision: a world where cars drive themselves, where innovation never stops, and where the way we move is constantly improving. Tesla's commitment to **innovation**, **sustainability**, and **autonomy** is reshaping the automotive industry and positioning it for a future unlike anything we've seen before. With each update, Tesla continues to lead the way toward a smarter, safer, and more efficient future of transportation. And as we look to the future, one thing is certain: Tesla will be at the forefront of this revolution, driving us toward a fully autonomous and electrified world.

www.ingramcontent.com/pod-product-compliance
Lightning Source LLC
Chambersburg PA
CBHW070421230526
45471CB00006B/2911